DESIGNING YOUR IDEAL MUSIC STUDIO

Unlocking Creativity, Optimizing Workflow, and Cultivating Inspiration

JERRY MARCUS

Copyright@2024

- CHAPTER ONE 5
  - INTRODUCTION TO MUSICAL STUDIO 5
- CHAPTER TWO 12
  - CHOOSE A SUITABLE LOCATION 12
- CHAPTER THREE 20
  - DESIGN LAYOUT 20
- CHAPTER FOUR 28
  - ACOUSTIC TREATMENT 28
- CHAPTER FIVE 37
  - SELECT AND INSTALL EQUIPMENTS 37
- CHAPTER SIX 50
  - SETUP RECORDING GEAR 50
- CHAPTER SEVEN 58
  - INSTALL MONITORING SYSTEM 58
- CHAPTER EIGHT 66
  - ARRANGE FUNITURES 66
- CHAPTER TEN 83
  - FINAL TESTING AND CALIBRATION 83
- CHAPTER ELEVEN 92
  - DECORATIONS 92
- CHAPTER TWELVES 101

DOCUMENTATION AND MAINTENANCE....101

CHAPTER THIRTEEN 111

  PROMOTIONS AND MARKETING 111

CHAPTER FOURTEEN......................................120

  CONCLUSION..120

CHAPTER ONE

INTRODUCTION TO MUSICAL STUDIO

Setting up a music studio is an exciting endeavor that requires careful planning, investment, and attention to detail.

Whether you're a budding musician, producer, or audio engineer, creating your own studio space provides the opportunity to unleash your creativity and produce high-quality recordings. In this introduction, we'll cover the essential steps to help you set up your own musical studio:

Define your goals and budget

Defining your goals and budget is the crucial first step in setting up a music studio. Here's how you can approach it:

Identify Your Goals:

Determine the primary purpose of your studio.

Are you planning to use it for personal projects, recording demos, producing professional tracks, mixing and mastering for clients, or a combination of these?

Clarify the types of music or audio content you intend to create.

For example, are you focusing on music production, podcasting, voice-over work, etc.?

Consider your long-term aspirations for the studio. Are you aiming to turn it into a full-time business, or is it primarily a hobbyist pursuit?

Establish Your Budget:

Assess your financial resources realistically. Determine how much you can comfortably invest in setting up and maintaining the studio.

Break down your budget into categories such as equipment, acoustic treatment, construction or renovation (if needed), furniture, utilities, and ongoing expenses.

Prioritize your spending based on your goals and the critical components needed to achieve them. Allocate more funds to areas that directly impact the quality of your studio's output.

Consider Additional Costs:

Account for any unforeseen expenses or hidden costs that may arise during the setup process.

Factor in ongoing expenses such as rent (if applicable), utilities, software subscriptions, equipment maintenance, and marketing (if you plan to offer services to clients).

Be prepared to adjust your budget as you gather more information about the specific requirements of your studio setup.

Seek Guidance if Necessary:

If you're unsure about setting a realistic budget or need advice on prioritizing expenses, consider consulting with professionals in the audio engineering or music production industry.

Research online forums, community groups, or industry publications to gather insights from experienced studio owners or professionals who have gone through similar setups.

By defining your goals and budget clearly, you can make informed decisions throughout the studio setup process and ensure that your investment aligns with your aspirations and financial capabilities.

Let's get started...

CHAPTER TWO
CHOOSE A SUITABLE LOCATION

Choosing a suitable location for your music studio is crucial for its success. Here's how to go about it:

Assess Your Needs:

Determine the size and layout requirements of your studio based on your goals and the type of activities you plan to undertake (e.g., recording, mixing, rehearsal).

Consider the amount of space needed for equipment, acoustic treatment, and comfortable workflow.

Consider Acoustics:

Look for spaces with good acoustic properties or the potential for improvement through acoustic treatment.

Avoid locations with excessive noise pollution from traffic, neighbors, or other sources that could interfere with recording sessions.

Accessibility:

Choose a location that is easily accessible for yourself and potential clients or collaborators.

Consider proximity to public transportation, parking availability, and any accessibility requirements for individuals with disabilities.

Evaluate Environmental Factors:

Ensure the space has adequate ventilation and climate control to maintain a comfortable working environment.

Check for potential issues such as moisture, mold, or pests that could affect equipment or health.

Budget Considerations:

Assess the cost of renting or purchasing a space in different locations and neighborhoods.

Balance your budgetary constraints with the desirability of the location in terms of convenience, amenities, and potential for attracting clients.

Legal and Zoning Requirements:

Research local zoning regulations and building codes to ensure that your chosen location is suitable for operating a music studio.

Obtain any necessary permits or licenses required for commercial use of the space.

Explore Different Options:

Consider various types of spaces, including commercial properties, live/work spaces, shared studios, or home-based setups.

Evaluate the pros and cons of each option in terms of cost, flexibility, privacy, and potential for growth.

Seek Input from Professionals:

Consult with real estate agents, property managers, or commercial brokers specializing in commercial properties to explore available options.

Get advice from experienced studio owners or audio professionals who can offer insights based on their own experiences with different locations.

Visit Potential Locations:

Arrange site visits to inspect potential locations firsthand and assess their suitability for your needs.

Pay attention to factors such as room dimensions, layout, natural light, and overall ambiance.

Make an Informed Decision:

Evaluate the pros and cons of each location based on your priorities and constraints.

Choose the location that best aligns with your budget, requirements, and long-term goals for your music studio.

By carefully considering these factors, you can select a suitable location that provides an optimal environment for your music studio and sets the stage for its success.

CHAPTER THREE
DESIGN LAYOUT

Designing the layout of your music studio is essential for creating an efficient and functional workspace. Here's how to approach it:

Assess Available Space:

Measure the dimensions of the room or space where you'll be setting up your studio.

Take note of any architectural features, such as windows, doors, or alcoves, that may impact the layout.

Identify Zones:

Divide the space into different zones based on the activities you'll be performing. Common zones include:

Recording area: where instruments or vocalists will be recorded.

Mixing and mastering area: where your computer workstation, monitors, and audio interface will be located.

Equipment storage: for instruments, microphones, cables, and other gear.

Comfort/lounge area: for relaxation and brainstorming sessions.

Consider Workflow:

Plan the layout to facilitate a smooth workflow from recording to mixing and mastering.

Position recording equipment and instruments in proximity to minimize cable runs and optimize signal flow.

Place mixing and mastering equipment within easy reach of your computer workstation.

Acoustic Considerations:

Take into account the acoustic properties of the room when designing the layout.

Position recording equipment and monitors to minimize reflections and standing waves.

Allocate space for acoustic treatment, such as bass traps, diffusers, and acoustic panels, to optimize sound quality.

Ergonomics and Comfort:

Choose furniture and equipment placement that promotes good posture and comfort during long recording and mixing sessions.

Ensure adequate lighting in all areas of the studio to reduce eye strain.

Incorporate seating areas or lounges where you and your collaborators can relax and recharge between sessions.

Cable Management:

Plan cable routes and storage solutions to minimize clutter and tripping hazards.

Use cable trays, raceways, or zip ties to organize cables neatly and prevent tangling.

Future Expansion:

Leave room for future expansion or upgrades as your studio grows.

Consider flexible furniture and equipment arrangements that can be easily reconfigured to accommodate changes in your workflow or equipment lineup.

Safety Considerations:

Ensure that all electrical equipment is installed and grounded properly to prevent electrical hazards.

Keep emergency exits clear and accessible at all times.

Visualization and Mockups:

Create a floor plan or 3D model of the studio layout to visualize the final design.

Use software tools or physical mockups to experiment with different layouts and configurations before finalizing your design.

Feedback and Iteration:

Seek feedback from colleagues, friends, or mentors on your proposed layout.

Be prepared to iterate on your design based on feedback and new insights that arise during the planning process.

By carefully designing the layout of your music studio, you can create a functional and inspiring workspace that enhances your creativity and productivity.

CHAPTER FOUR
ACOUSTIC TREATMENT

Acoustic treatment is essential for creating a controlled and balanced sound environment in your music studio. Here's how to approach it:

Assess the Acoustic Properties:

Evaluate the existing acoustic characteristics of the room, including its size, shape, and construction materials.

Identify any acoustic issues such as flutter echoes, standing waves, or excessive reverb that need to be addressed.

Determine Treatment Goals:

Define the acoustic goals for your studio, such as achieving a neutral listening environment for mixing and mastering or minimizing external noise for recording sessions.

Consider the primary activities in the studio (recording, mixing, mastering) and tailor the treatment to support those activities.

Select Treatment Materials:

Choose appropriate acoustic treatment materials based on your goals and budget. Common options include:

Acoustic panels:

Absorb sound reflections to reduce reverberation and control room resonance. Panels are available in various sizes, thicknesses, and designs (e.g., fabric-wrapped, foam, fiberglass).

Bass traps: Absorb low-frequency sound waves to minimize bass buildup and improve clarity in the low end. Bass traps are typically placed in corners where low-frequency energy tends to accumulate.

Diffusers: Scatter sound reflections to create a more natural and spacious acoustic environment. Diffusers are useful for breaking up strong reflections without overly deadening the room.

Ceiling clouds: Suspended panels or baffles installed on the ceiling to absorb sound reflections and reduce overhead flutter echoes.

Placement and Coverage:

Determine the optimal placement of acoustic treatment panels and other materials based on the room's acoustic characteristics and your treatment goals.

Aim for even coverage throughout the room to achieve a balanced frequency response and consistent sound quality.

Pay particular attention to reflection points where sound waves bounce off walls, ceiling, and floor, as well as corners where bass energy tends to accumulate.

Room Mode Analysis:

Use room mode calculators or acoustic measurement tools to identify and address standing wave frequencies that may cause uneven bass response.

Place bass traps strategically in locations where room modes are most pronounced to mitigate their effects on the listening environment.

Installation:

Install acoustic treatment materials securely according to the manufacturer's recommendations.

Use appropriate mounting hardware and techniques to ensure stability and effectiveness of the treatment.

Test and Adjust:

Conduct acoustic measurements or listening tests to evaluate the effectiveness of the treatment.

Make adjustments as needed to fine-tune the room's acoustic properties and achieve the desired sound quality.

Monitor and Maintain:

Regularly monitor the condition of acoustic treatment materials and make repairs or replacements as needed.

Keep the room clean and free of dust to maintain the effectiveness of the treatment over time.

By implementing effective acoustic treatment in your music studio, you can create a conducive environment for recording, mixing, and mastering, allowing you to achieve accurate and professional-quality audio productions.

CHAPTER FIVE

SELECT AND INSTALL EQUIPMENTS

Selecting the right equipment is crucial for the success of your music studio. Here's a guide to help you choose the essential gear:

Recording Interface:

Choose an audio interface that suits your recording needs and budget.

Consider the number of inputs and outputs required for simultaneous recording of instruments and microphones.

Look for features such as preamps, phantom power, and compatibility with your recording software.

Microphones:

Invest in a selection of microphones suitable for various recording applications.

Include dynamic, condenser, and ribbon microphones to cover different sound sources and recording scenarios.

Research microphone characteristics like frequency response, polar pattern, and sensitivity to choose the right mic for each situation.

Studio Monitors:

Select high-quality studio monitors with flat frequency response and accurate sound reproduction.

Consider factors like driver size, amplifier power, and room acoustics when choosing monitor speakers.

Budget for monitor isolation pads or stands to decouple the speakers from the desk or floor and minimize vibrations.

Headphones:

Invest in professional studio headphones for critical listening and monitoring during recording, mixing, and mastering.

Choose closed-back headphones for tracking and open-back headphones for mixing and mastering to achieve accurate sound representation.

Look for headphones with comfortable padding and adjustable headbands for extended use.

Mixer (Optional):

Consider adding a mixer to your setup for routing and controlling multiple audio sources simultaneously.

Choose a mixer with enough channels and routing options to accommodate your recording and mixing needs.

Look for features like built-in effects, EQ, and dynamics processing if desired.

Digital Audio Workstation (DAW) Software:

Select a DAW software platform that suits your workflow and preferences.

Consider factors like ease of use, compatibility with your operating system, and availability of essential features such as recording, editing, mixing, and mastering tools.

Explore different DAW options and trial versions before making a final decision.

Cables, Stands, and Accessories:

Budget for high-quality cables and connectors to ensure reliable signal transmission between your audio equipment.

Purchase microphone stands, boom arms, and shock mounts to position microphones securely and minimize handling noise.

Invest in accessories like pop filters, windshields, and cable management solutions to improve the efficiency and professionalism of your studio setup.

Optional Gear:

Consider additional equipment based on your specific recording and production needs, such as:

MIDI controllers for virtual instruments and software synths.

Outboard gear like compressors, equalizers, and effects processors for analog signal processing.

Instruments, amplifiers, and other sound sources to expand your creative palette.

Budget Allocation:

Allocate your budget based on the essential equipment needed to achieve your recording and production goals.

Prioritize investments in critical components like audio interfaces, microphones, and studio monitors, and allocate remaining funds to accessories and optional gear.

Research and Demo:

Research equipment options online, read reviews, and watch demo videos to familiarize yourself with different brands and models.

Visit music stores or attend trade shows to demo equipment in person and assess its suitability for your studio setup.

By carefully selecting the right equipment for your music studio, you can create a professional and versatile workspace that meets your recording and production needs.

CHAPTER SIX
SETUP RECORDING GEAR

Setting up recording gear in your music studio involves connecting and configuring various components to facilitate recording sessions effectively.

Here's a step-by-step guide to help you through the process:

Prepare the Room:

Ensure that the recording space is clean, organized, and free from clutter.

Set up acoustic treatment panels and bass traps to optimize the room's acoustics and minimize unwanted reflections.

Position Microphones:

Place microphones in appropriate positions relative to the sound source(s) you intend to record.

Experiment with microphone placement to achieve the desired tone and balance.

Use microphone stands, booms, or mounts to position microphones securely and minimize handling noise.

Connect Microphones to Audio Interface:

Use XLR cables to connect microphones to the inputs of your audio interface.

Ensure that each microphone is connected to the correct input channel on the interface.

Set Input Levels:

Adjust the input gain on your audio interface to set appropriate levels for each microphone.

Aim for a healthy signal level without clipping or distortion, typically indicated by LED meters on the interface.

Configure Monitoring Setup:

Connect studio monitors or headphones to the outputs of your audio interface.

Adjust the monitor volume to a comfortable listening level for tracking and recording.

Configure Recording Software (DAW):

Launch your digital audio workstation (DAW) software on your computer.

Create a new project/session and configure audio settings such as sample rate, bit depth, and input/output routing.

Assign input channels in the DAW to correspond with the inputs on your audio interface.

Arm Tracks for Recording:

Arm audio tracks in your DAW for recording, specifying the input source (microphone) for each track.

Set recording levels and enable monitoring to listen to incoming audio through headphones or studio monitors.

Perform Soundchecks:

Conduct soundchecks to ensure that microphone placement, input levels, and monitor settings are optimal for recording.

Record test takes to listen back and make adjustments as needed before starting formal recording sessions.

Record Takes:

Start recording takes once you've confirmed that everything is set up correctly and the performers are ready.

Monitor recording levels during the session to prevent clipping or distortion.

Monitor and Review Takes:

Listen back to recorded takes to evaluate performance quality and technical aspects of the recording.

Make note of any issues or areas for improvement and address them as necessary before proceeding.

Save and Backup Recording Files:

Save recording sessions in your DAW project folder and create backups to prevent data loss.

Organize files systematically and label them accurately for easy retrieval and future reference.

Maintain and Troubleshoot:

Regularly check and maintain recording gear to ensure optimal performance.

Troubleshoot any technical issues promptly to minimize downtime during recording sessions.

By following these steps, you can set up recording gear in your music studio efficiently and create an environment conducive to high-quality recording sessions.

CHAPTER SEVEN

INSTALL MONITORING SYSTEM

Installing a monitoring system is essential for accurate playback and critical listening in your music studio.

Here's a step-by-step guide to help you install and configure your monitoring system effectively:

Positioning Studio Monitors:

Place studio monitors on sturdy stands or speaker mounts at ear level.

Position monitors symmetrically and equidistant from the listening position to create an accurate stereo image.

Angle monitors slightly inward towards the listening position to create a focused listening area.

Room Acoustics Consideration:

Take into account the acoustic properties of the room when positioning studio monitors.

Use acoustic treatment such as bass traps, acoustic panels, and diffusers to minimize unwanted reflections and optimize the listening environment.

Connecting Studio Monitors:

Connect the studio monitors to the appropriate outputs on your audio interface or monitor controller using balanced audio cables (e.g., XLR or TRS cables).

Ensure that cables are securely connected and free from interference.

Calibrating Monitor Placement:

Use a tape measure and laser level to ensure precise positioning of studio monitors.

Measure and mark the distance from each monitor to the listening position to ensure equal distance for balanced stereo imaging.

Setting Monitor Levels:

Adjust the volume controls on the studio monitors to a comfortable listening level for critical listening.

Use the monitor level control on your audio interface or monitor controller to fine-tune the overall volume level.

Room Calibration (Optional):

Consider using room calibration software or hardware to analyze and correct room acoustics and monitor frequency response.

Measure room modes and frequency response using measurement microphones and software tools to identify areas for improvement.

Testing Monitor Placement:

Play reference tracks with a known frequency response to evaluate the accuracy of your monitor placement.

Listen for any anomalies in frequency response, stereo imaging, and overall sound balance.

Optimizing Listening Position:

Position the listening chair or couch at the optimal listening position, typically equidistant from the studio monitors and centered between them.

Adjust the height and angle of the listening position to achieve the best balance between direct sound and room reflections.

Testing and Fine-Tuning:

Listen to a variety of audio material across different genres and styles to assess the performance of your monitoring system.

Make adjustments to monitor placement, room treatment, and listening position as needed to optimize sound quality and accuracy.

Regular Monitoring Maintenance:

Clean studio monitors regularly to remove dust and debris that may affect performance.

Check and tighten connections periodically to prevent signal degradation or intermittent issues.

By following these steps, you can install and configure a monitoring system that provides accurate playback and critical listening capabilities in your music studio, enabling you to produce high-quality audio recordings and mixes.

CHAPTER EIGHT

ARRANGE FUNITURES

Arranging furniture and ensuring ergonomic comfort in your music studio is crucial for maintaining productivity and reducing the risk of fatigue or injury during long recording and mixing sessions. Here's how to approach it:

Assess Space and Workflow:

Evaluate the available space in your studio and consider the workflow and activities that will take place.

Determine the optimal placement of furniture and equipment to facilitate an efficient and ergonomic workspace.

Select Ergonomic Furniture:

Choose ergonomic studio furniture that promotes comfort and supports good posture.

Invest in an adjustable studio chair with lumbar support, adjustable armrests, and seat height to accommodate different users and preferences.

Select a studio desk or workstation with an ergonomic design that provides ample space for equipment and allows for easy reach of essential tools and controls.

Position Studio Equipment:

Arrange recording and mixing equipment such as audio interfaces, keyboards, and control surfaces within easy reach of the primary workstation.

Ensure that monitors, keyboards, and other essential tools are positioned at the correct height and angle to reduce strain on the neck, shoulders, and wrists.

Consider Monitor Placement:

Position studio monitors at ear level and at an equal distance from the listening position to create an accurate stereo image.

Use monitor isolation pads or stands to decouple monitors from the desk or floor and minimize vibrations.

Create Comfortable Listening Areas:

Designate comfortable seating areas or lounges where you and your collaborators can relax and recharge between recording or mixing sessions.

Include amenities such as coffee tables, couches, and bean bags to create a welcoming and collaborative atmosphere.

Organize Cables and Accessories:

Implement cable management solutions to keep cables tidy and prevent tripping hazards.

Use cable trays, raceways, or zip ties to bundle and route cables neatly behind furniture and equipment.

Adjust Lighting:

Ensure adequate lighting in the studio to reduce eye strain and create a comfortable working environment.

Install task lighting such as adjustable desk lamps or overhead lights with dimming controls to customize the lighting levels according to the task at hand.

Optimize Room Acoustics:

Position acoustic treatment panels, bass traps, and diffusers strategically to optimize room acoustics and minimize unwanted reflections.

Balance absorption and diffusion to create a balanced and natural acoustic environment conducive to critical listening and accurate monitoring.

Test and Adjust:

Sit in various positions within the studio to assess comfort and functionality.

Make adjustments to furniture placement, monitor height, and seating arrangements based on personal preferences and ergonomic principles.

Regular Maintenance:

Periodically review and adjust furniture and equipment arrangements to accommodate changes in workflow or equipment setup.

Clean and maintain furniture regularly to ensure longevity and optimal performance.

By prioritizing ergonomic comfort and efficient furniture arrangement in your music studio, you can create a conducive workspace that promotes creativity, productivity, and well-being during recording, mixing, and production activities.

CHAPTER NINE

POWER AND ELECTRICAL CONSIDERATIONS

Power and electrical considerations are critical for ensuring the safety and reliability of your music studio setup. Here's a comprehensive guide:

Assess Power Requirements:

Determine the total power consumption of all equipment in your studio, including audio interfaces, monitors, amplifiers, and outboard gear.

Check the power ratings (in watts or amps) of each device and add them up to calculate the total power demand.

Evaluate Existing Electrical Infrastructure:

Inspect the electrical system in your studio space, including outlets, circuit breakers, and wiring.

Determine the capacity of your electrical service (e.g., 100 amps, 200 amps) and identify any limitations or deficiencies.

Plan Outlet Placement:

Map out the locations of power outlets needed for each piece of equipment in your studio.

Ensure that outlets are conveniently located near workstations and equipment racks to minimize the need for extension cords and power strips.

Consider Dedicated Circuits:

Allocate dedicated circuits for sensitive audio equipment to prevent interference and ensure stable power supply.

Install separate circuits for high-power devices like amplifiers or studio monitors to avoid overloading existing circuits.

Install Ground Fault Circuit Interrupters (GFCIs):

Install GFCI outlets in areas where water or moisture is present, such as near sinks or in bathrooms, to protect against electrical shocks.

Consider installing GFCI breakers for additional protection on circuit panels.

Protect Against Power Surges:

Install surge protectors or uninterruptible power supplies (UPS) to safeguard audio equipment from power spikes and voltage fluctuations.

Use surge-protected power strips or outlets for sensitive electronics like computers and audio interfaces.

Ensure Proper Grounding:

Verify that all electrical outlets and equipment are properly grounded to minimize the risk of electrical shock and interference.

Consult a qualified electrician if you encounter grounding issues or are unsure about the grounding status of your studio setup.

Label Circuit Breakers:

Label circuit breakers in the electrical panel to clearly identify which circuits correspond to specific areas or equipment in your studio.

Create a detailed electrical plan or diagram indicating the location of outlets, circuits, and breaker assignments for reference.

Adhere to Building Codes and Regulations:

Familiarize yourself with local building codes and regulations governing electrical installations in commercial or residential spaces.

Ensure that your studio setup complies with safety standards and requirements to prevent accidents and regulatory issues.

Regular Maintenance and Inspection:

Schedule periodic inspections of electrical systems and equipment to identify and address any potential hazards or deficiencies.

Routinely check power cords, outlets, and connections for signs of wear or damage, and replace as needed. By addressing power and electrical considerations proactively, you can create a safe and reliable environment for your music studio setup, minimizing the risk of electrical hazards and ensuring uninterrupted operation of your audio equipment.

CHAPTER TEN

FINAL TESTING AND CALIBRATION

Final testing and calibration are crucial steps to ensure that your music studio is optimized for accurate audio reproduction and professional-quality recordings.

Here's a step-by-step guide to help you through the process:

Room Acoustic Measurement:

Use measurement microphones and room analysis software to measure the acoustic properties of your studio.

Conduct frequency response measurements and analyze room modes to identify areas of concern that may affect sound quality.

Monitor Calibration:

Use monitor calibration software or hardware to measure and adjust the frequency response of your studio monitors.

Generate correction curves based on room measurements to compensate for any anomalies in frequency response.

Speaker Placement Optimization:

Fine-tune the placement of studio monitors based on acoustic measurements and calibration results.

Adjust monitor positioning to minimize standing waves, reflections, and other room-induced anomalies.

Listening Environment Evaluation:

Listen to reference tracks across different genres and styles to evaluate the accuracy and balance of your monitoring system.

Pay attention to details such as frequency balance, stereo imaging, and depth perception to assess the overall sound quality.

Critical Listening Sessions:

Conduct critical listening sessions to evaluate the performance of your studio setup in real-world scenarios.

Listen to recorded tracks and mixes to assess tonal balance, dynamics, and spatial imaging.

Make notes of any areas for improvement or adjustments needed in the setup.

Equipment Performance Check:

Test the performance of audio interfaces, microphones, and other recording gear to ensure proper functionality.

Check for signal integrity, noise levels, and overall sound quality during recording and playback.

Calibration Verification:

Verify the effectiveness of monitor calibration and room correction settings by comparing reference tracks with and without correction applied.

Make adjustments as needed to fine-tune the calibration settings for optimal results.

Documentation and Record Keeping:

Document the results of room measurements, monitor calibration, and critical listening evaluations for future reference.

Keep a log of any adjustments made to equipment settings or room treatment to track changes over time.

Final Tweaks and Adjustments:

Make final tweaks and adjustments to optimize the studio setup based on feedback from critical listening sessions and performance tests.

Experiment with different monitor positions, acoustic treatment configurations, and equipment settings to achieve the desired sound quality.

Re-test and Verify:

Re-test the studio setup periodically to ensure that it maintains optimal performance and accuracy over time.

Verify calibration settings and room acoustics regularly to address any changes or drift that may occur.

By conducting thorough testing and calibration of your music studio, you can fine-tune the setup to achieve accurate audio reproduction and professional-quality recordings, providing a conducive environment for creativity and productivity.

CHAPTER ELEVEN
DECORATIONS

Decorating and personalizing your music studio can enhance its ambiance, inspire creativity, and make it feel like a welcoming and inspiring space. Here are some ideas to consider:

Choose a Theme or Style:

Decide on a theme or style that reflects your personality, interests, and the type of music you create.

Whether it's minimalist, vintage, modern, or eclectic, choose decor elements that resonate with your aesthetic preferences.

Display Artwork and Memorabilia:

Hang artwork, posters, or album covers on the walls to add visual interest and inspiration.

Showcase memorabilia such as concert tickets, vinyl records, or musical instruments that hold sentimental value or represent milestones in your music career.

Customize with Lighting:

Install ambient lighting fixtures or LED strips to create mood lighting and set the tone for creative sessions.

Use task lighting such as adjustable desk lamps or spotlights to illuminate specific work areas and provide focused illumination.

Incorporate Plants and Greenery:

Bring nature indoors by adding plants and greenery to your studio space.

Choose low-maintenance plants like succulents, ferns, or snake plants to add freshness and improve air quality.

Accessorize with Textiles:

Add warmth and texture to your studio with decorative textiles such as rugs, curtains, or throw pillows.

Choose fabrics with sound-absorbing properties to improve acoustics and reduce echo in the room.

Create a Comfortable Lounge Area:

Designate a cozy lounge area with comfortable seating where you and your collaborators can relax and brainstorm ideas.

Add plush couches, bean bags, or floor cushions for informal gatherings and listening sessions.

Install Inspirational Quotes or Lyrics:

Display motivational quotes, lyrics, or affirmations on wall decals or chalkboards to inspire creativity and uplift spirits.

Choose quotes that resonate with your musical aspirations and philosophy.

DIY Projects and Upcycling:

Get creative with DIY projects or upcycling to personalize your studio on a budget.

Repurpose old furniture or musical instruments into unique decor pieces, or create custom artwork using recycled materials.

Functional Decor Elements:

Incorporate decor elements that serve a functional purpose, such as shelving units for storage or display, or acoustic panels with decorative fabric covers.

Choose furniture pieces that combine style with practicality, such as studio desks with built-in cable management or multi-purpose storage ottomans.

Rotate and Refresh:

Keep your studio decor fresh and inspiring by rotating artwork, rearranging furniture, or updating accessories periodically.

Experiment with seasonal decorations or themed accents to infuse new energy into the space throughout the year.

By decorating and personalizing your music studio, you can create an inviting and inspiring environment that nurtures creativity, enhances productivity, and reflects your unique style and identity as a musician or producer.

CHAPTER TWELVES

DOCUMENTATION AND MAINTENANCE

Documentation and maintenance are essential aspects of managing a music studio effectively, ensuring that equipment remains in optimal condition and operations run smoothly. Here's a comprehensive guide:

Document Equipment Inventory:

Create a detailed inventory of all equipment and gear in your music studio, including serial numbers, purchase dates, and warranty information.

Maintain an updated list of software licenses, plugins, and other digital assets.

Create Maintenance Schedules:

Develop maintenance schedules for all equipment based on manufacturer recommendations and usage patterns.

Schedule routine maintenance tasks such as cleaning, lubricating moving parts, and replacing consumable items (e.g., cables, microphone capsules).

Establish Equipment Maintenance Procedures:

Document step-by-step procedures for performing routine maintenance tasks on each piece of equipment.

Include safety precautions, recommended tools, and troubleshooting tips in maintenance procedures.

Track Repairs and Service Records:

Keep records of all repairs and servicing performed on equipment, including dates, issues addressed, and service provider information.

Document any modifications or upgrades made to equipment over time.

Backup Data Regularly:

Implement a regular backup schedule for all digital data stored in your music studio, including project files, samples, and recordings.

Use redundant backup solutions such as external hard drives, cloud storage, or network-attached storage (NAS) to protect against data loss.

Label and Organize Cables:

Label all cables and connectors in your studio setup to facilitate troubleshooting and maintenance.

Use cable management solutions such as cable ties, sleeves, and trays to keep cables organized and prevent tangling.

Monitor Room Environment:

Monitor environmental factors such as temperature and humidity levels in your studio to ensure optimal conditions for equipment and sound quality.

Use digital hygrometers and thermometers to track environmental conditions and address any fluctuations that may occur.

Document Studio Layout and Wiring:

Create a diagram or schematic of your studio layout, including equipment placement, signal flow, and wiring connections.

Keep records of any changes or updates made to the studio setup over time.

Train Staff and Collaborators:

Provide training to staff and collaborators on proper equipment usage, maintenance procedures, and studio protocols.

Encourage team members to report any issues or concerns promptly and provide feedback on ways to improve studio operations.

Regular Inspections and Audits:

Conduct regular inspections of equipment, wiring, and studio infrastructure to identify potential issues or hazards.

Perform periodic audits of inventory, maintenance logs, and documentation to ensure accuracy and completeness.

Plan for Equipment Upgrades:

Develop a plan for equipment upgrades and replacements based on technological advancements, changing workflow requirements, and budget considerations.

Research new gear and technologies regularly to stay informed about industry trends and innovations.

By implementing comprehensive documentation and maintenance practices in your music studio, you can minimize downtime, extend the lifespan of equipment, and ensure consistent performance and reliability for your recording and production activities.

CHAPTER THIRTEEN

PROMOTIONS AND MARKETING

Promotion and marketing are crucial for attracting clients, building your brand, and growing your music studio business. Here's a guide to help you develop effective promotion and marketing strategies:

Define Your Brand Identity:

Identify your unique selling points and what sets your music studio apart from competitors.

Develop a brand identity that reflects your values, personality, and target audience.

Create a Professional Website:

Design and launch a professional website that showcases your music studio, services, and portfolio.

Include high-quality photos, videos, client testimonials, and pricing information to attract potential clients.

Utilize Social Media Platforms:

Establish a presence on social media platforms such as Facebook, Instagram, Twitter, and LinkedIn.

Share updates, behind-the-scenes content, client testimonials, and promotions to engage with your audience and attract new clients.

Build Relationships with Local Businesses:

Network with local businesses, music schools, recording artists, and event organizers to promote your music studio.

Offer partnership opportunities, discounts, or referral programs to incentivize collaborations and word-of-mouth referrals.

Offer Special Promotions and Packages:

Create special promotions, discounts, or package deals to attract new clients and incentivize repeat business.

Consider offering introductory rates, referral bonuses, or loyalty rewards to encourage client retention.

Attend Industry Events and Networking Mixers:

Attend music industry events, conferences, and networking mixers to connect with potential clients and industry professionals.

Distribute business cards, flyers, and promotional materials to raise awareness of your music studio.

Create Compelling Content:

Produce informative and engaging content such as blog posts, videos, podcasts, or tutorials related to music production, recording techniques, and studio equipment.

Share content on your website, social media channels, and email newsletters to position yourself as an authority in your field and attract organic traffic.

Optimize for Search Engines (SEO):

Optimize your website and online content for search engines to improve visibility and attract organic traffic.

Use relevant keywords, meta tags, and descriptions to optimize your website's ranking in search engine results pages (SERPs).

Collect and Showcase Client Testimonials:

Collect testimonials and reviews from satisfied clients and showcase them on your website and social media channels.

Positive testimonials and word-of-mouth recommendations can build trust and credibility with potential clients.

Invest in Online Advertising:

Consider investing in online advertising channels such as Google Ads, Facebook Ads, or Instagram Ads to reach a broader audience and promote your music studio services.

Set clear objectives, target demographics, and budgets for your advertising campaigns to maximize ROI.

Track and Analyze Results:

Monitor and analyze the performance of your promotion and marketing efforts using analytics tools.

Track key metrics such as website traffic, social media engagement, leads generated, and conversion rates to measure the effectiveness of your strategies.

By implementing a comprehensive promotion and marketing plan, you can increase awareness of your music studio, attract new clients, and grow your business in a competitive industry.

CHAPTER FOURTEEN

CONCLUSION

Continuous learning and improvement are essential for staying relevant, adapting to industry changes, and refining your skills as a music studio owner.

Here are some strategies to foster ongoing growth and development:

Stay Updated with Industry Trends:

Keep abreast of the latest trends, technologies, and developments in the music production and recording industry.

Follow industry publications, blogs, forums, and social media channels to stay informed about emerging trends, software updates, and new gear releases.

Attend Workshops and Training Sessions:

Participate in workshops, seminars, and training sessions related to music production, audio engineering, and studio management.

Look for opportunities to learn from industry experts, attend masterclasses, and acquire new skills to enhance your expertise.

Invest in Continuing Education:

Enroll in courses, online classes, or certification programs to deepen your knowledge and skills in specific areas of music production or studio management.

Explore topics such as mixing and mastering techniques, music theory, digital audio workstations (DAWs), and sound design.

Practice Regularly:

Dedicate time to practice your craft regularly, whether it's experimenting with new recording techniques, refining your mixing skills, or mastering new software tools.

Set aside time for creative exploration and experimentation to push your boundaries and discover new approaches to music production.

Seek Feedback and Mentorship:

Solicit feedback from peers, mentors, and industry professionals to gain valuable insights and perspectives on your work.

Surround yourself with a supportive network of fellow musicians, engineers, and producers who can provide constructive criticism and guidance.

Reflect and Evaluate:

Take time to reflect on your successes and failures, and evaluate what worked well and what could be improved in your music studio operations.

Use feedback and performance metrics to identify areas for growth and set actionable goals for continuous improvement.

Experiment with New Techniques and Tools:

Embrace a spirit of experimentation and innovation by exploring new recording techniques, production methods, and audio processing tools.

Stay open to trying out different software plugins, hardware gear, and creative workflows to expand your repertoire and evolve your sound.

Network and Collaborate:

Build relationships with fellow musicians, producers, and industry professionals through networking events, online communities, and collaborative projects.

Collaborate on music production projects, remixes, or co-writing sessions to gain new perspectives and learn from others' experiences.

Stay Organized and Efficient:

Implement efficient workflows, organizational systems, and time management techniques to streamline your music studio operations.

Use project management tools, scheduling apps, and task lists to stay organized and prioritize tasks effectively.

Stay Inspired and Passionate:

Cultivate a sense of curiosity, passion, and enthusiasm for music production and recording.

Stay inspired by exploring diverse musical genres, attending live performances, and immersing yourself in creative activities outside of your studio.

By embracing a mindset of continuous learning and improvement, you can stay ahead of the curve, adapt to evolving industry trends, and elevate your skills as a music studio owner and audio professional.

www.ingramcontent.com/pod-product-compliance
Lightning Source LLC
Chambersburg PA
CBHW050306230526
45471CB00005B/2044